THE ANGULAR RATIO
TRANSFERENCE (ART) HYPOTHESES

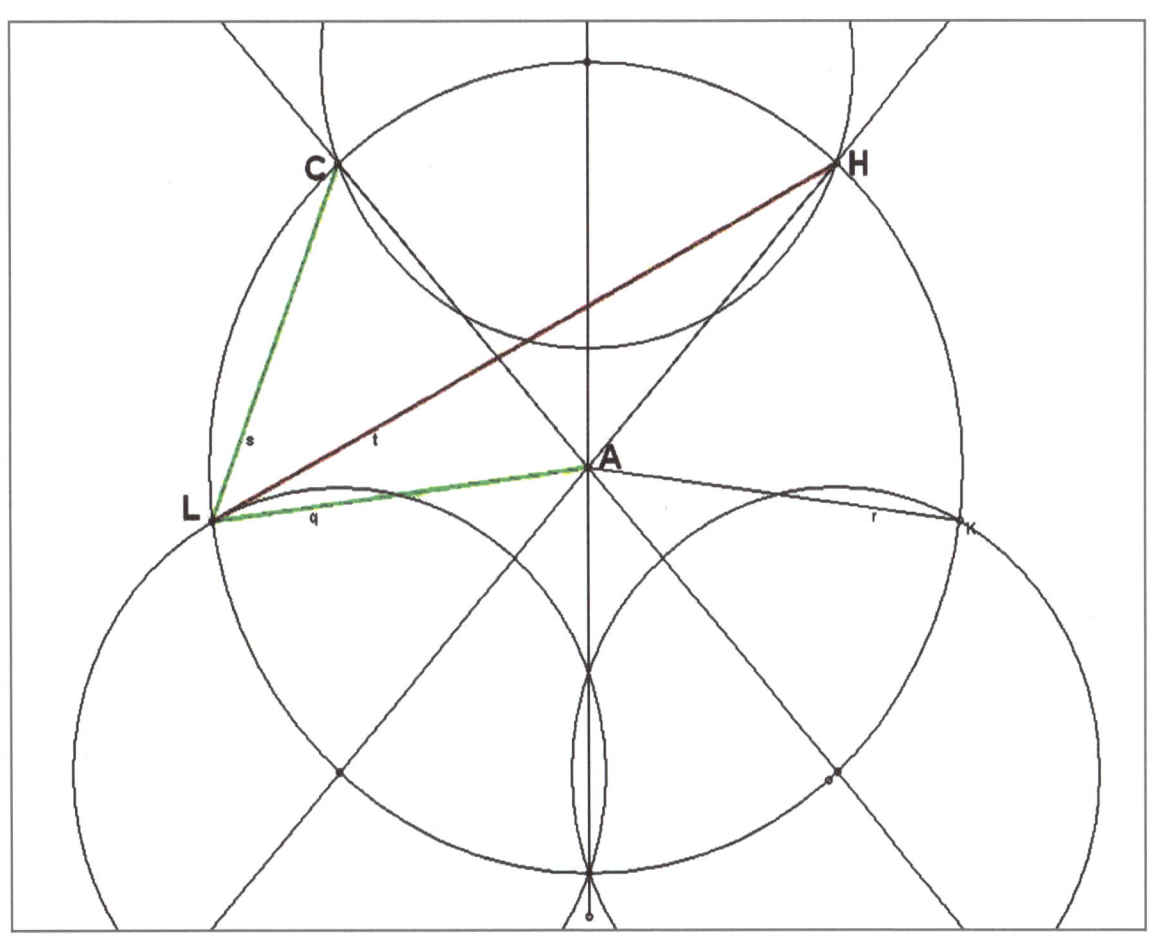

A Fresh Approach to an Ancient Problem in Plane Geometry

James Mulberry

To order additional copies of this book, contact:
Xlibris Corporation
1-888-795-4274
www.Xlibris.com
Orders@Xlibris.com

Here is the paradox

Any angle from 0° through 90° generated from the construction steps described herein can be shown to be trisectable under Euclid's ground rules. However, **we aren't yet able to start** with the given angle between 0° and 90° and trisect it directly.

Therein lie the paradox!

I. Introduction

The Angular Ratio Transference (ART) hypotheses contends that it is possible to transfer the bisector of a reference angle to another angle in such an original way that the bisector of the reference angle is transformed into the trisector of the other angle. This ratio transformation from 1/2 to 1/3rd is made possible by the unique relationship of the two angles within a random circle.

To facilitate the understanding of the ART process this paper is divided into six sections: (1) introduction, (2) the construction methodology, (3) the general ART equation (4) examples, (5) conclusion, and (6) references.

This paper presents a fresh approach to an intriguing problem in plane geometry. There are a multitude of articles and websites dedicated to Euclid's problem children: squaring a circle, doubling a cube, and trisecting a given angle. However, we will only focus on just one of them: the trisection of a given angle. For most people knowledgeable in the history of mathematics, the solvability or insolvability of these problems has already been put to rest by such mathematicians as François Viète (1540-1603) and Pierre Wantzel—in his seminal work: "'Recherches sur les moyens de reconnaître si un Problème de Géométrie peut se résoudre avec la règle et le compas."(. . . . *a method to ascertain whether a geometric problem can be solved by ruler and compass*—1836). A typical representative quote on the impossibility of angular trisection is: "These three problems drew attention of mathematicians for more than 2,000 years. The problems were never solved geometrically because with only a straightedge and compass they ***cannot*** be solved. That is an entirely different statement from saying that the solution has not been found yet. **The solution was not found because it does not exist**. This remarkable fact was discovered by using a new and very powerful type of algebra developed during the 19th century", Ph. D. John Tabak, *The History of Mathematics (2004), Geometry – The Language of Space and Form*, page 21. A mathematical argument runs like this: "More generally, "constructible" numbers are those that are algebraic of order a power of two—that is, that they satisfy an irreducible polynomial equation with integer coefficients of degree a power of 2 (1, 2, 4, 8, etc.). $\sqrt{n}$ satisfies the equation $x^2-n=0$ and so is either algebraic of order 2 (if x^2-n is irreducible—cannot be factored with integer coefficients) or algebraic of order 1: both of which are powers of two. We can construct a segment of length $n^{1/m}$ if and only if m is a power of two."

"That, by the way, is the reason we can't trisect all angles. cos(20 degrees) satisfies the irreducible polynomial equation $8x^3-6x-1=0$: it is algebraic of order 3 and so is not "constructible". It's easy to construct a 60 degree angle: give an unit length, strike arcs of that length from each end point and draw the line from one end point to a point of intersection of those arcs. **IF** it were possible to trisect that 60 degree angle, it would be possible to construct a 20 degree angle. Now strike of a unit length on one leg of that angle and drop a perpendicular to the other. You would have constructed a segment of length cos(20) which is impossible." (PF Mentor—http://www.physicsforums.com/showthread.php?t=163146, march 17, 2008)

Non Euclidean Solutions

Many solutions to the trisection challenge have been proposed over the 2500 years since Euclid first introduced his problem children and none have survived the test of objective analysis and repeatability. Very close approximations have been achieved by some ingenious approaches, however, the only exact solutions that have been achieved have been through the application of non Euclidean techniques. Some of those approaches include: (1) the use of special curves like Nicomedes' conchoid (~200 b.c.), the Quadratrix (see figure below), and later the trisectrix (see figure below), (2) the use of mechanical instruments such a T-square devices shaped like a tomahawk, (3) the use of origami (paper folding), and (4) the use of mathematical converging formulas.

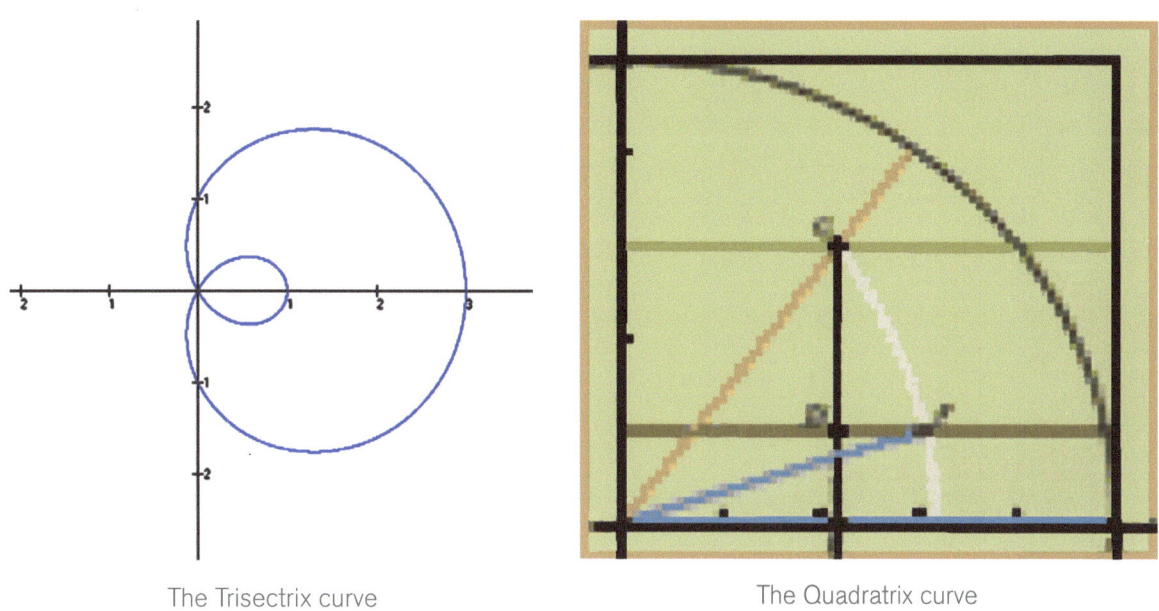

The Trisectrix curve   The Quadratrix curve

The ART process (which this paper is about) is a breakthrough for angular trisection. The process demonstrates that all acute angles are trisectable, but it does not yet take the final step—starting with a given angle—which is the author's main objective. That final step is still being researched. Also, it is not the purpose of this paper to rehash the history, failed attempts, and the mathematical arguments about the insolubility of these problems. That will be left to the reader's own initiative, however, an abbreviated list of references is provided for those who may wish to research this subject further.

The ART process is like entering a house by going in through the back door. It's a paradigm shift in the way we approach a problem. In other words, rather than working directly on a given angle in order to trisect it, the ART process first begins with a circle within which an acute angle will be created. It's the unique relationship of the inner angles to the circle that allows one to remove the historical barriers to this geometry problem and to trisect the resultant angle.

The hint that the impossible might just be possible is illustrated in the construction of a circle. Certain numbers or ratios can be constructed under Euclid's ground rules regardless of the restrictions imposed by high order mathematics. For instance, take the transcendental number pi (= 3.14159 . . .). Mathematicians tell us that pi cannot be drawn or constructed with an unmarked straightedge and compass. "Lindemann proved it in the 1890s that [pi] is not the root of any polynomial. The proof is hard. It is just possible to follow it if you had a very good two semester class in Calculus and were at the top of the class. But there is no doubt about it, [pi] is not the root of any polynomial, and every constructible number is such a root, so [pi] cannot be constructed by Euclidean moves. But squaring the circle, as we showed above, amounts to such a construction of [pi]. Therefore the circle cannot be squared"—selfAdjoint—http://physicsforums.com/archive/index.php/t-4359.html).

Impossible constructions

The algebraic characterization of constructible numbers provides an important *necessary* condition for constructibility: if z is constructible, then it is algebraic, and its minimal irreducible polynomial has degree and power of 2, or equivalently, the field extension $Q(z)/Q$ has dimension and power of 2. One should note that it is true, (but not obvious to show) that the converse is false — this is not a *sufficient* condition for constructibility. However, this defect can be remedied by considering the normal closure of $Q(z)/Q$.

The non-constructibility of certain numbers proves the impossibility of certain problems attempted by the philosophers of ancient Greece. In the following chart, each row represents a specific ancient construction problem. The left column gives the name of the problem. The second column gives an equivalent algebraic formulation of the problem. In other words, the solution to the problem is affirmative if and only if each number in the given set of numbers is constructible. Finally, the last column provides the simplest known counterexample. In other words, the number in the last column is an element of the set in the same row, but is not constructible.

Construction problem	Associated set of numbers	Counterexample
Doubling the cube	$\left\{ \sqrt[3]{x} : x \text{ is constructible} \right\}$	$\sqrt[3]{2}$ is not constructible, because its minimal polynomial has degree 3 over Q
Trisecting the angle	$\left\{ \cos\left(\dfrac{\arccos x}{3} \right) : x \text{ is constructible} \right\}$	$\cos\left(\dfrac{\arccos(1/2)}{3} \right) = \dfrac{1}{2}\left(2\cos\left(\dfrac{\pi}{9} \right) \right)$ is not constructible, because $2\cos\left(\dfrac{\pi}{9} \right)$ has minimal polynomial of degree 3 over Q
Squaring the circle	$\left\{ \sqrt{\pi} \right\}$	$\sqrt{\pi}$ is not constructible, because $\left(\sqrt{\pi} \right)^2 = \pi$ is not algebraic over Q
Constructing all regular polygons	$\left\{ e^{2\pi i/n} : n \in \mathbb{N}, n \geq 3 \right\}$	$e^{2\pi i/7}$ is not constructible, because 7 is not a Fermat prime

Yet we unconsciously utilize this transcendental number (or ratio) each time we draw a circle or diameter. Pi is an inherent part of a circumference (C = • D) and a diameter (D = C /). The unique relationship between a radius and its circumference allows one to complete a circle. If we could not include pi in these constructions we would have to redefine circle constructions and rewrite the textbooks on plane geometry.

II. Construction Methodology

The following nine construction steps illustrate the ART methodology and how each step can be drawn with an unmarked straightedge and compass in Euclidean space (in a flat plane). The following figures were drawn using a software application called Geometer's Sketchpad (version 3.10) and the measurement precision was set at 1/1000 of a degree. In addition, there is a rounding off algorithm imbedded within this program that may affect the last place digit. This 'rounding off', however, does not affect the correctness of the logic, the construction steps, or the associated equations.

Step 1
With a compass draw a circle of some random radius r

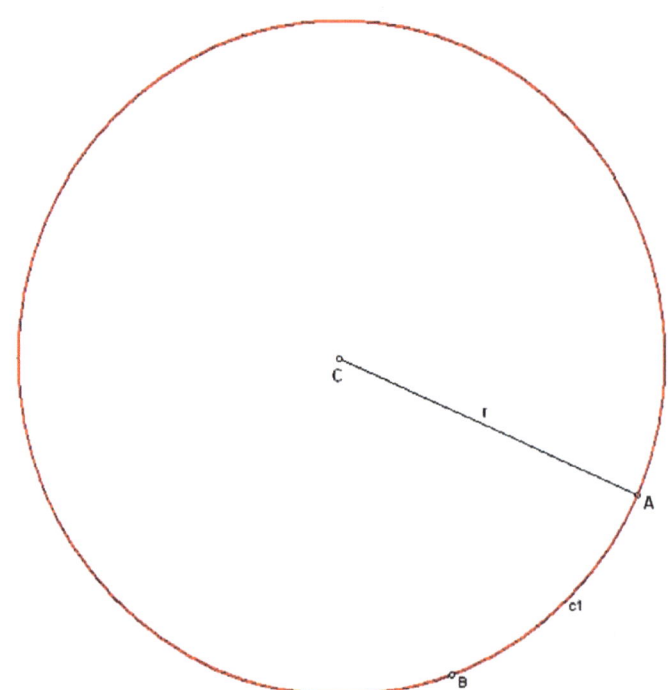

Step 2
With an
unmarked
straightedge
draw a diameter
BD

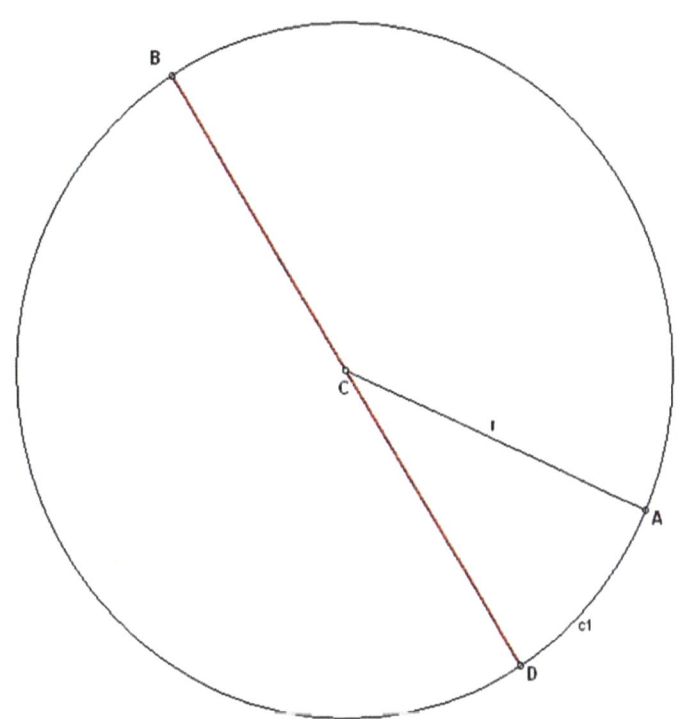

Step 3
With an
unmarked
straightedge
draw a diameter
EF

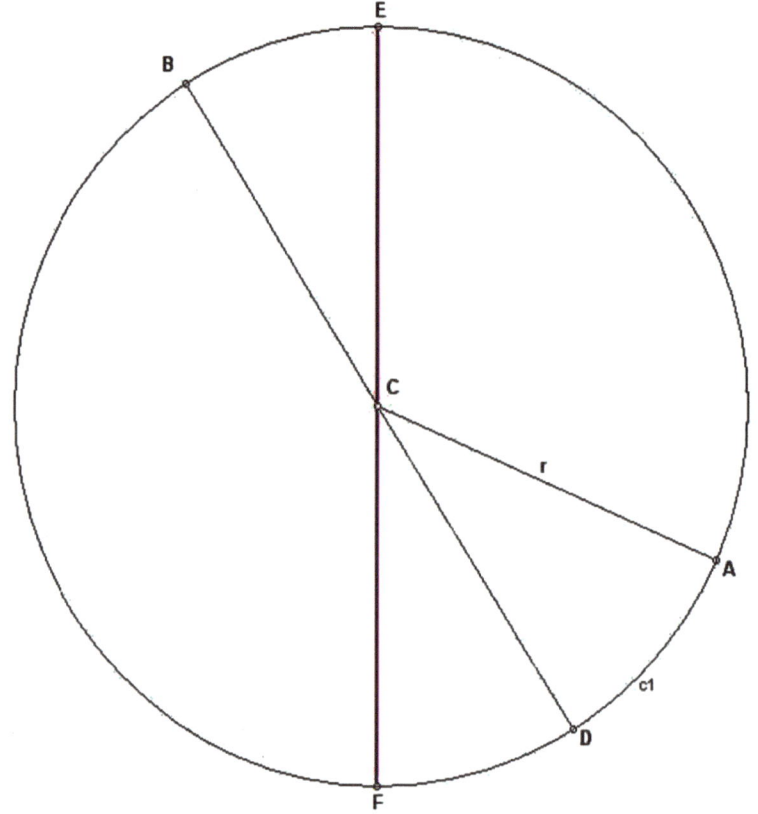

Step 4
With a compass on E as the center and radius EB draw a circle and mark its intersection with circle C as point G

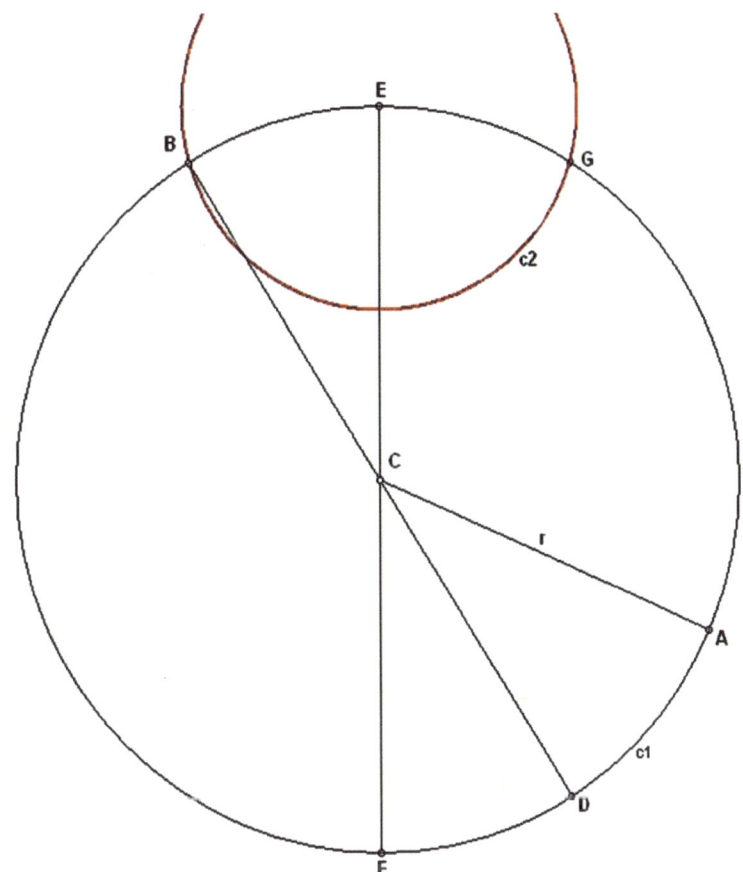

Step 5
With an unmarked
straightedge draw
diameter GH .

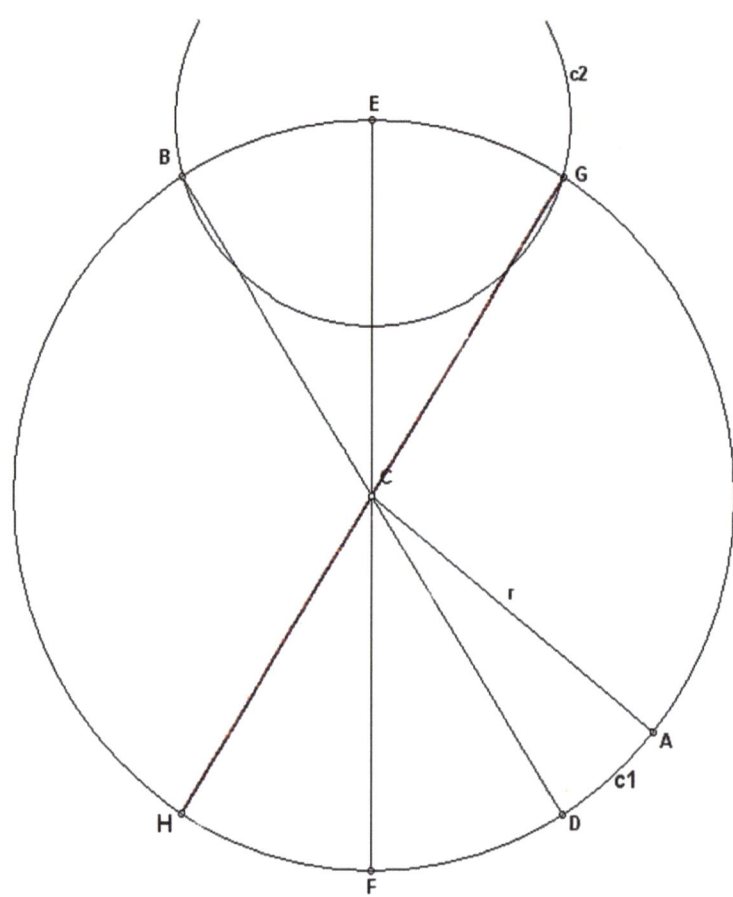

Step 6
With a compass at points H and D as centers and distances FH and FD as radii draw circles c3 and c4 and mark their intersection with circle C1 as points I and J.

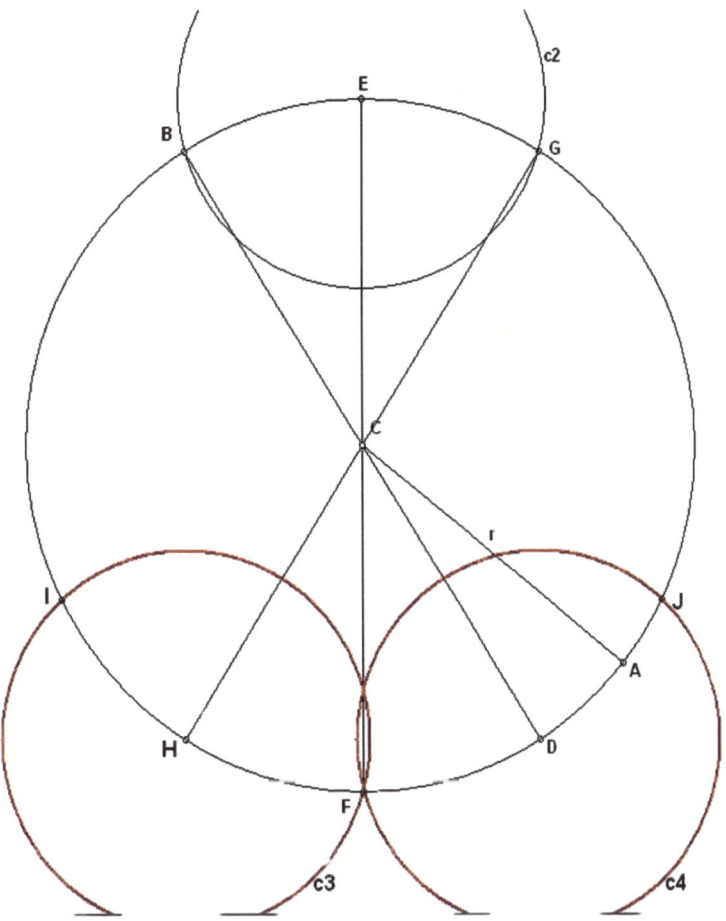

Step 7
With an unmarked
straightedge draw
radii IC and JC.

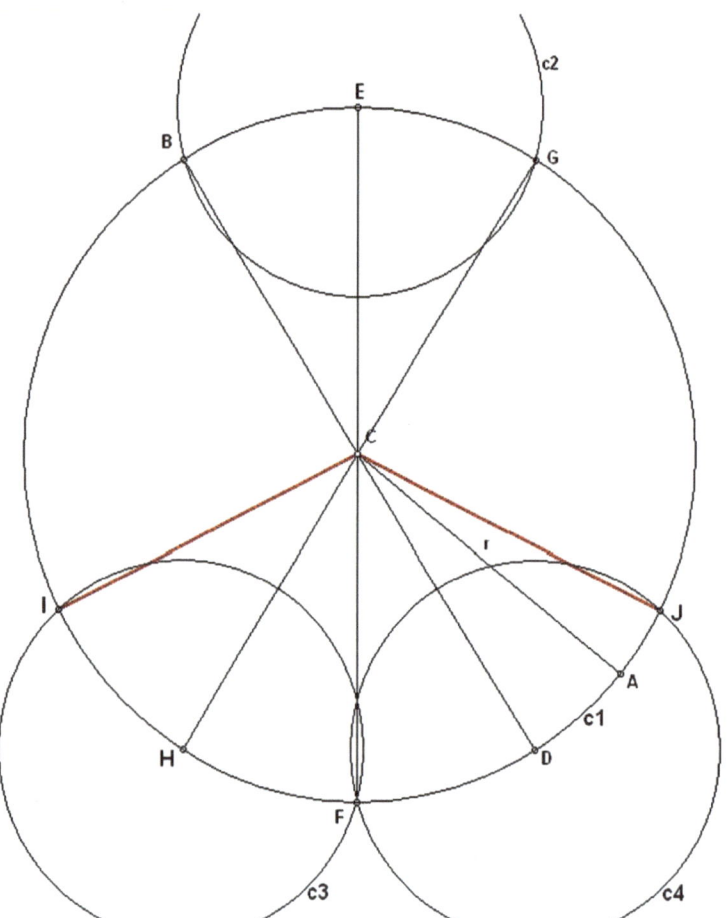

Let's examine what we have constructed so far. We have just created an ∠ ICJ that is directly opposite and twice the size of ∠ BCG. **These are the most important characteristics of the ART process**. If ∠ ICJ is of any other ratio or orientation to ∠ BCG then angular trisection cannot be obtained.

Step 8
With an unmarked straightedge draw line segments IB and JG.

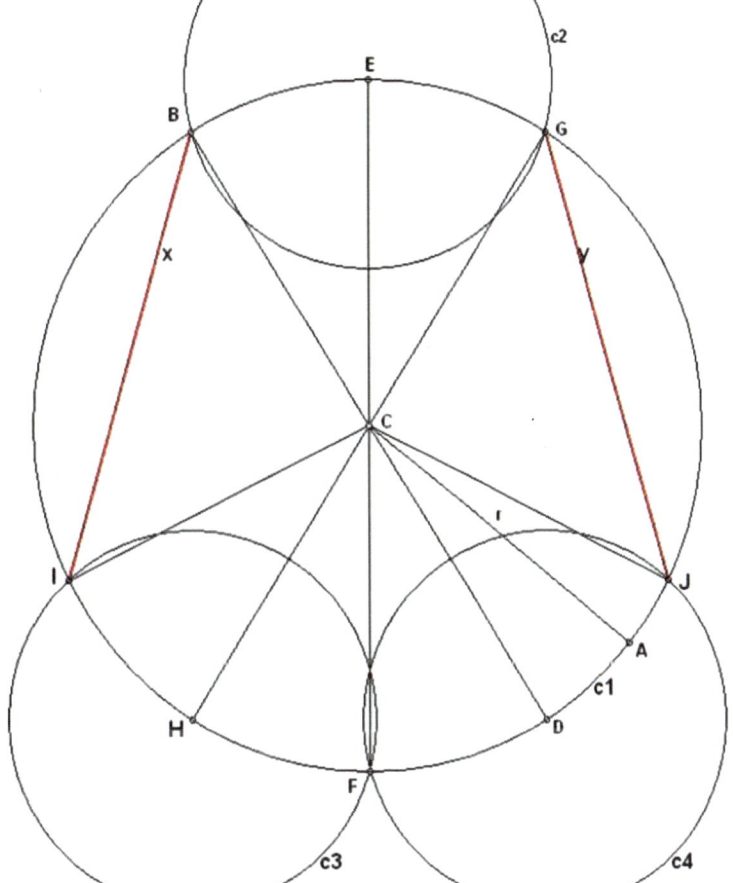

This step creates the acute ∠ BIC which will be the inscribed angle to be trisected. This is a good place to reiterate the point that we did not start out with a given angle. We have, however, generated a unique figure wherein we can create a random acute angle (the angular size determined solely by the selection of random points **B** and **E**. Remember the paradox stated on page 1: "*The purpose of this paper is to show that acute angles from 0° to 90° are trisectable under Euclid's ground rules using only an unmarked straightedge and a collapsible compass. It is not the purpose of this paper, however, to show the trisection process **starting** with a given angle.*"

Step 9
With an unmarked
straightedge draw
line segment IG.

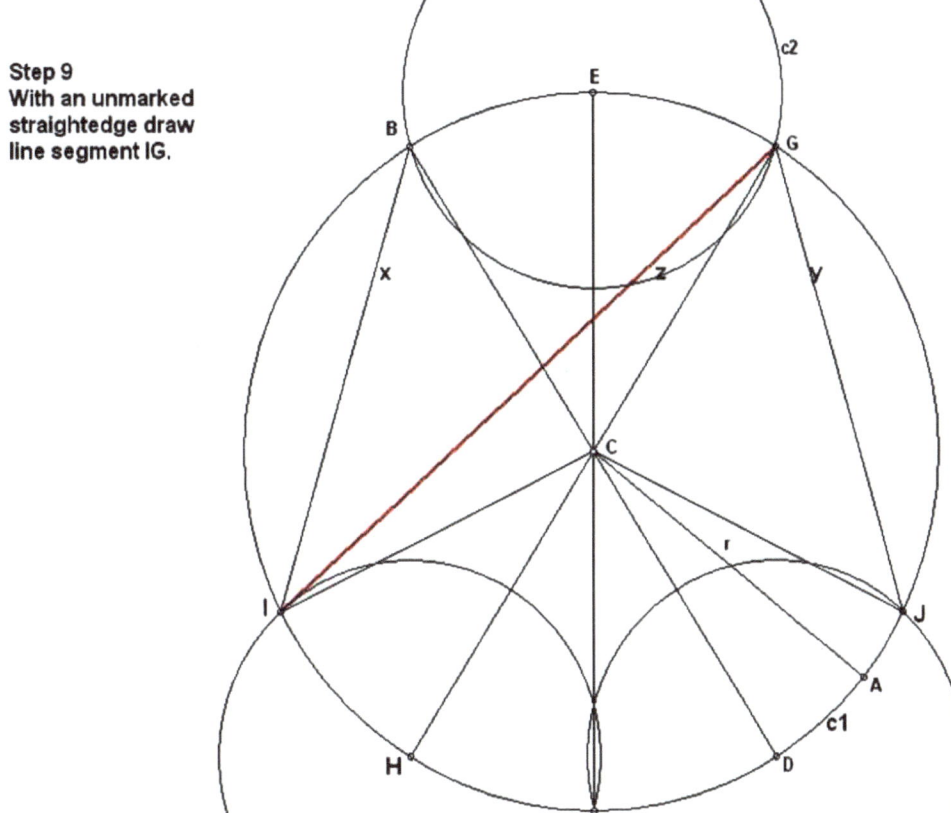

It is the author's contention that line (IG), drawn above in step nine, trisects ∠ CIB. This

construction technique works for any angle from 0° to 90. The size of ∠ BIC is driven only by

the selection of points **B** and **E.** All other construction steps fall out from the selection of these

two random points. For angles equal to or greater than 90° we first subtract and trisect all 90°

components using well known construction steps, and then, using the ART process, show that the

remaining acute angle is trisectable.

III. The General Trisection Equation

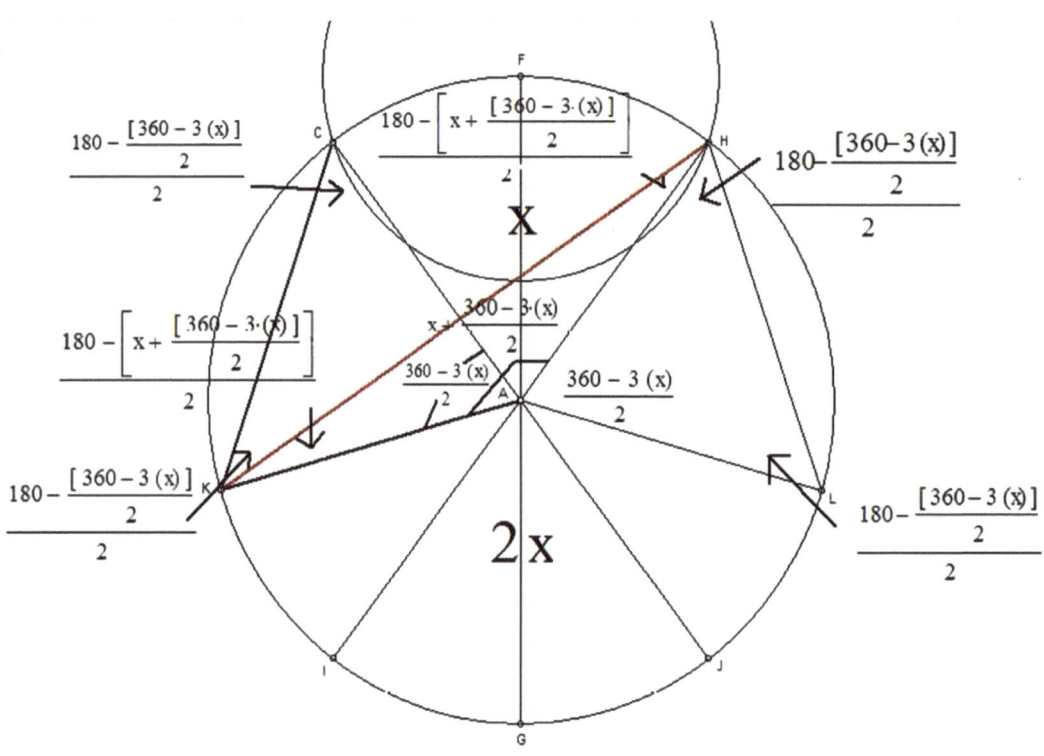

A diagram illustrating a general ratio transference with it's associated terms.

Above are the terms of the ART equation. We will examine each term with respect to the above figure:

1. **X** = a general ∠ CAH determined by the points C and F. By construction, ∠ CAF is one half ∠ CAH. See construction steps 2 through 5 above.

2. **2X** = an angle, by construction, that is twice the size and directly opposite of ∠ CAH. See steps 6 and 7.

3. ∠ s HAL and CAK = (360°−3x)/2

4.   ∠ s AKC, KCA, AHL, and HLA are all equal because they are base angles of two congruent isosceles triangles (AKC and ALH), which means their angles = (180°−(360°−3**X**) /2) /2.

5.   Triangle AKH is an isosceles triangle and ∠ KAH = **X** + (360−3(**X**)) / 2.

5.   ∠ s AKH and AHK are base angles of the isosceles triangle AHK . They are equal to 180°−(**X** + (360−3(**X**)) / 2) / 2.

The above terms lead to the following general ART equation:

$$\frac{\text{The size of } \angle \text{ AKH}}{\text{The size of } \angle \text{ AKC}} = \frac{1}{3} \text{ or}$$

$$\frac{180-\left[(x)+\dfrac{[\,360-3(x)\,]}{2}\right]}{2} \Bigg/ \frac{\left[180-\dfrac{[\,360-3(x)\,]}{2}\right]}{2} = \frac{1}{3} \text{⅂}$$

or .33333....⊙

For any **X** (within the range where ∠ CKA is from 0° to 90°)**,** the ratio of ∠ AKH to ∠ AKC is always and exactly .3333 . . . . .

IV. Examples

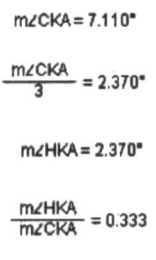

$$m\angle CKA = 7.110°$$

$$\frac{m\angle CKA}{3} = 2.370°$$

$$m\angle HKA = 2.370°$$

$$\frac{m\angle HKA}{m\angle CKA} = 0.333$$

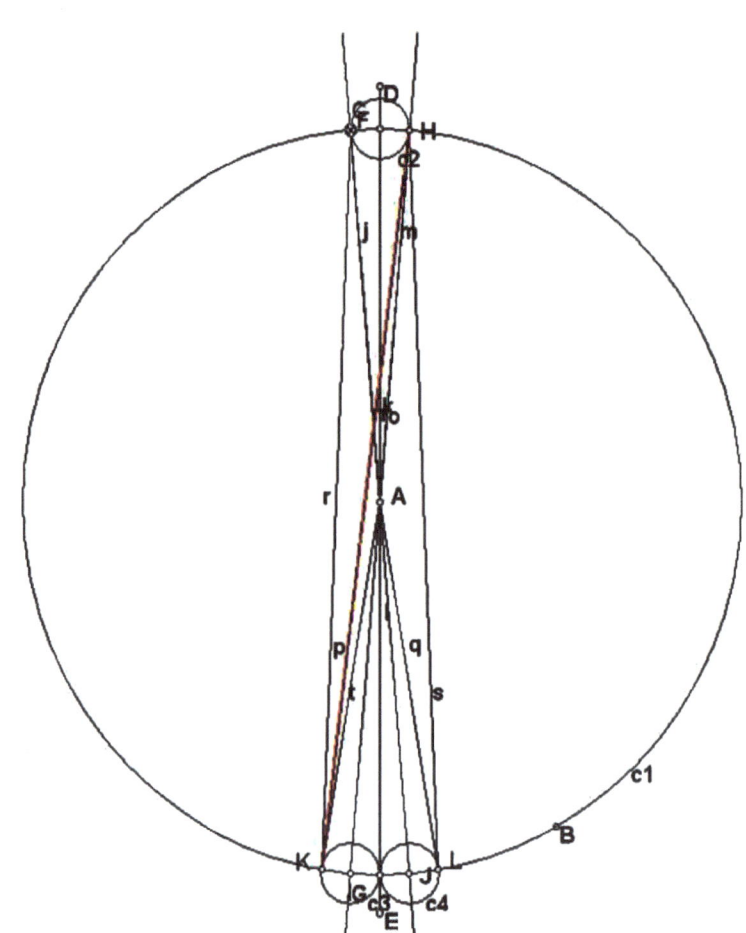

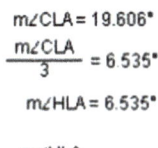
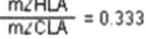

m∠CLA = 19.606°

$\frac{m\angle CLA}{3} = 6.535°$

m∠HLA = 6.535°

$\frac{m\angle HLA}{m\angle CLA} = 0.333$

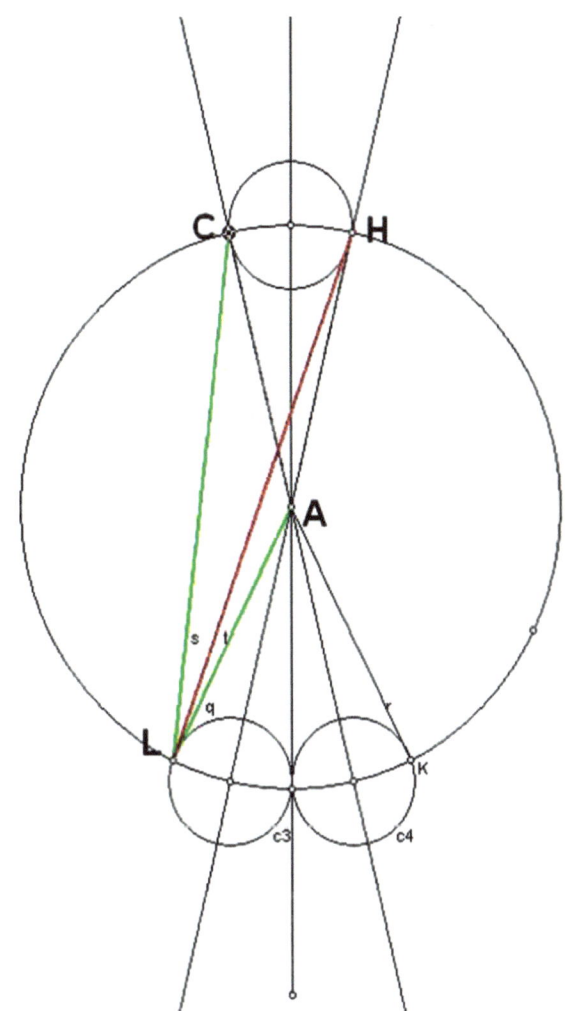

$m\angle CKA = 25.888°$

$\dfrac{m\angle CKA}{3} = 8.629°$

$m\angle HKA = 8.629°$

$\dfrac{m\angle HKA}{m\angle CKA} = 0.333$

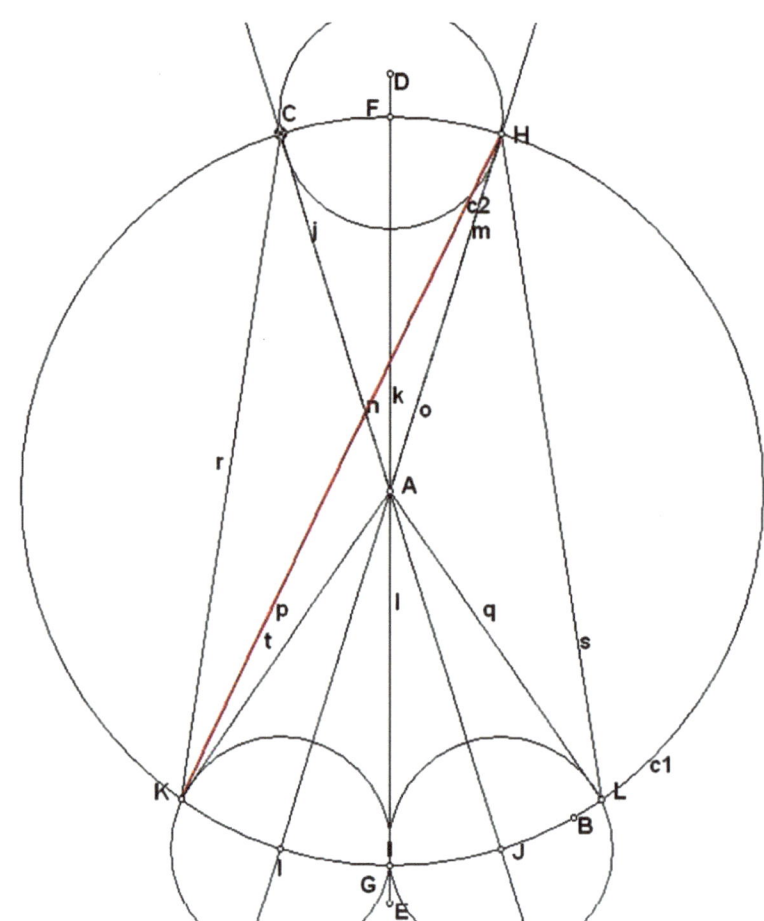

m∠CKA = 33.489°

$$\frac{m\angle CKA}{3} = 11.163°$$

m∠HKA = 11.163°

$$\frac{m\angle HKA}{m\angle CKA} = 0.333$$

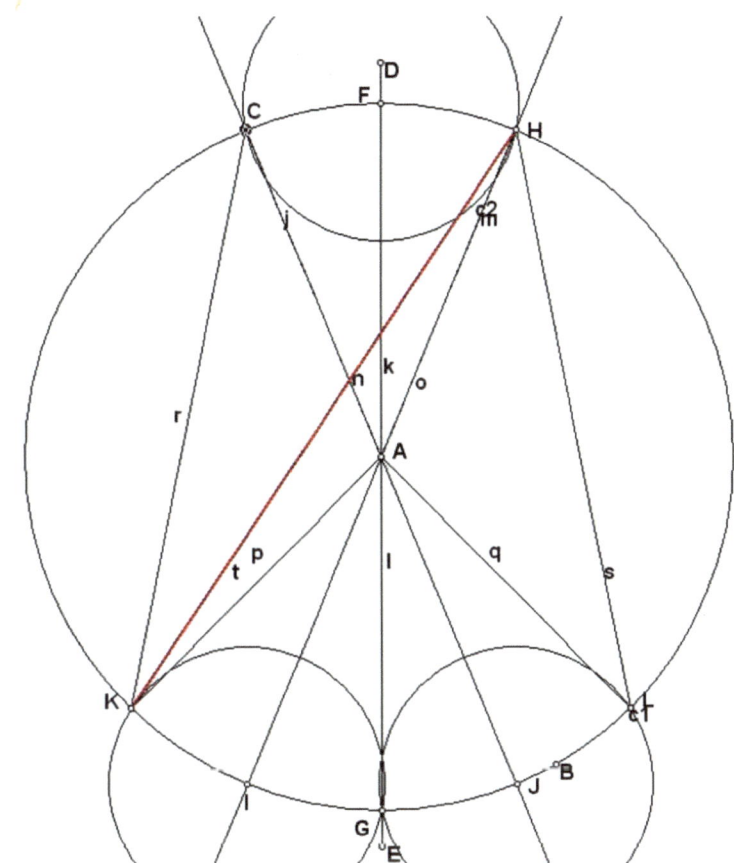

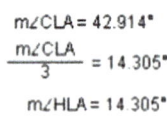

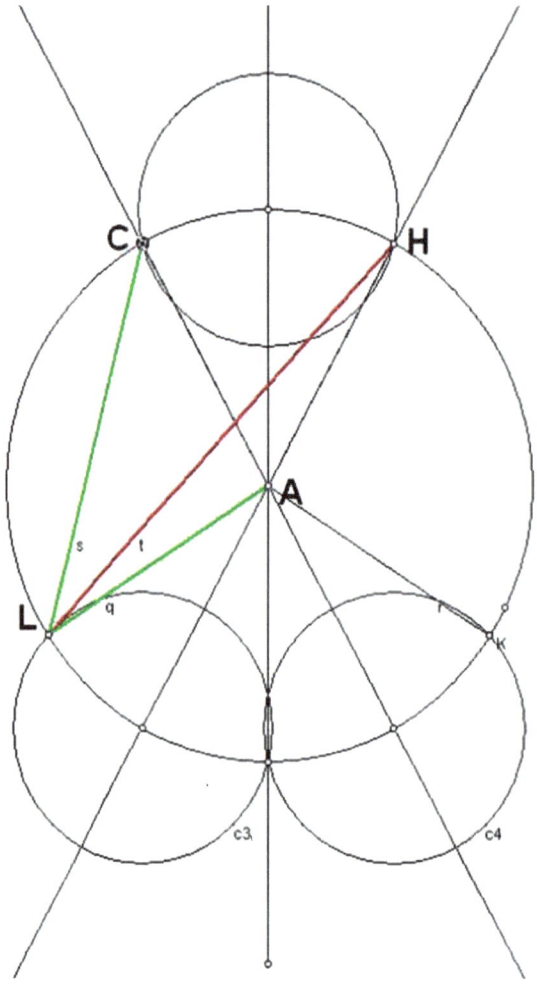

$m\angle CKA = 51.868°$

$\dfrac{m\angle CKA}{3} = 17.289°$

$m\angle HKA = 17.289°$

$\dfrac{m\angle HKA}{m\angle CKA} = 0.333$

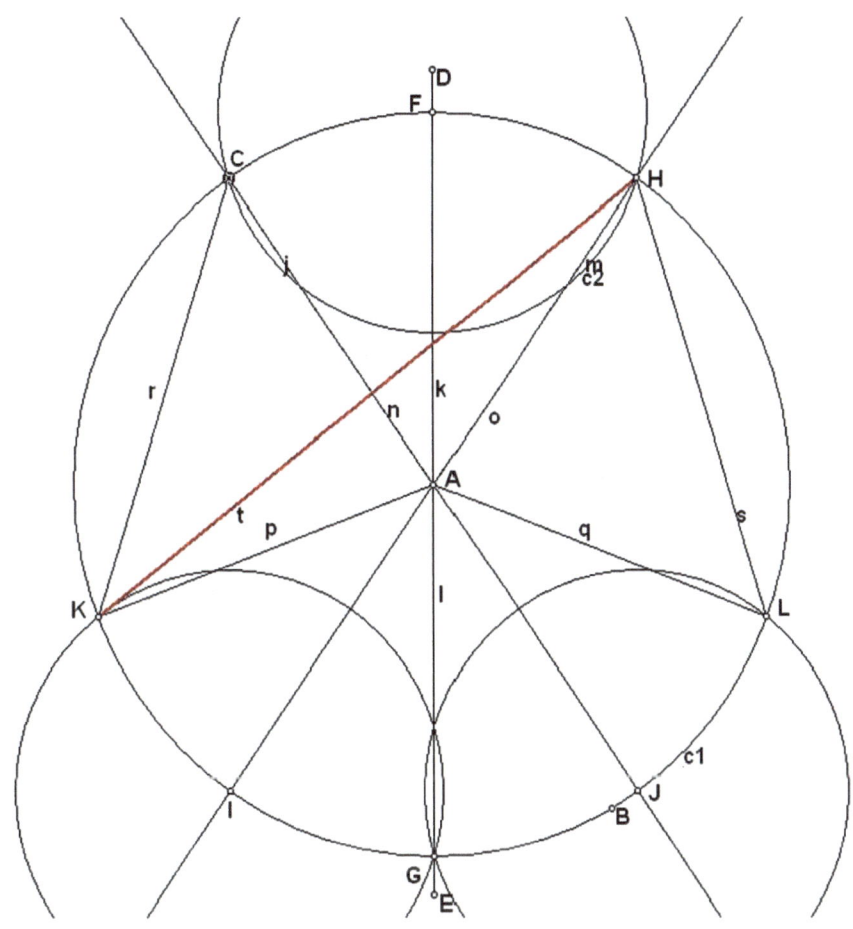

$m\angle CLA = 60.035°$

$\dfrac{m\angle CLA}{3} = 20.012°$

$m\angle HLA = 20.012°$

$\dfrac{m\angle HLA}{m\angle CLA} = 0.333$

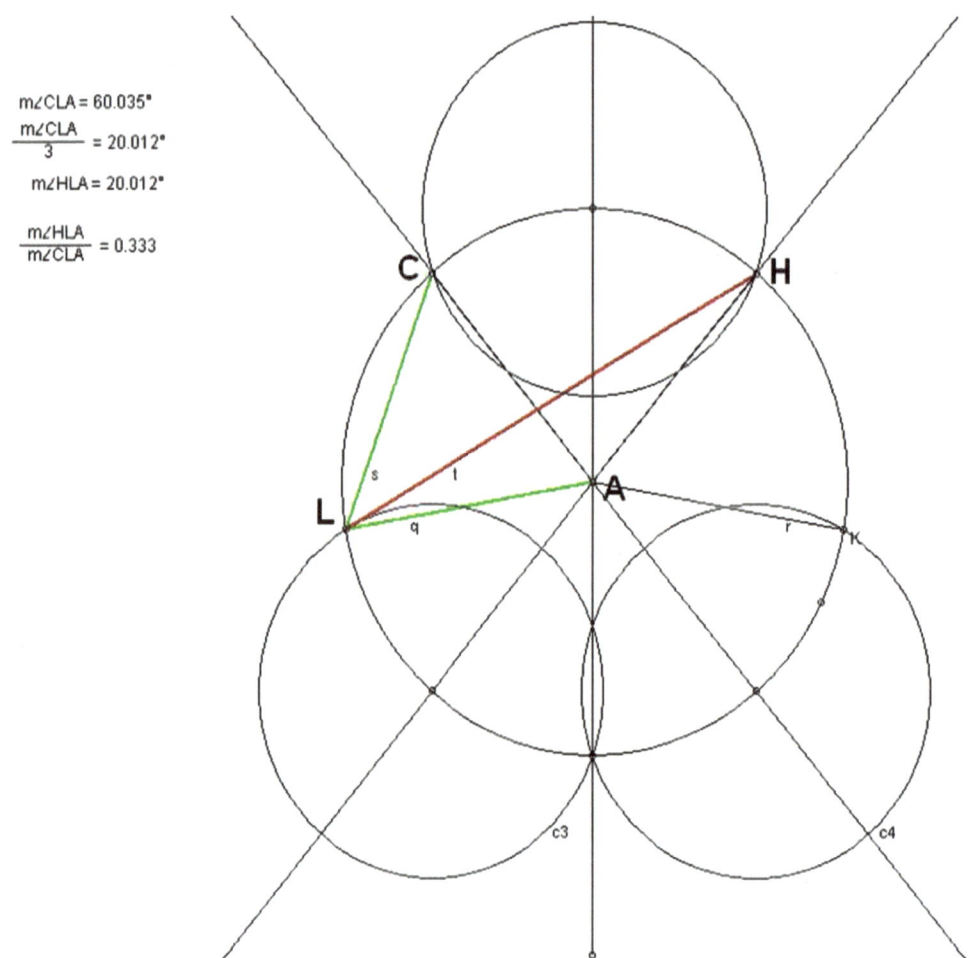

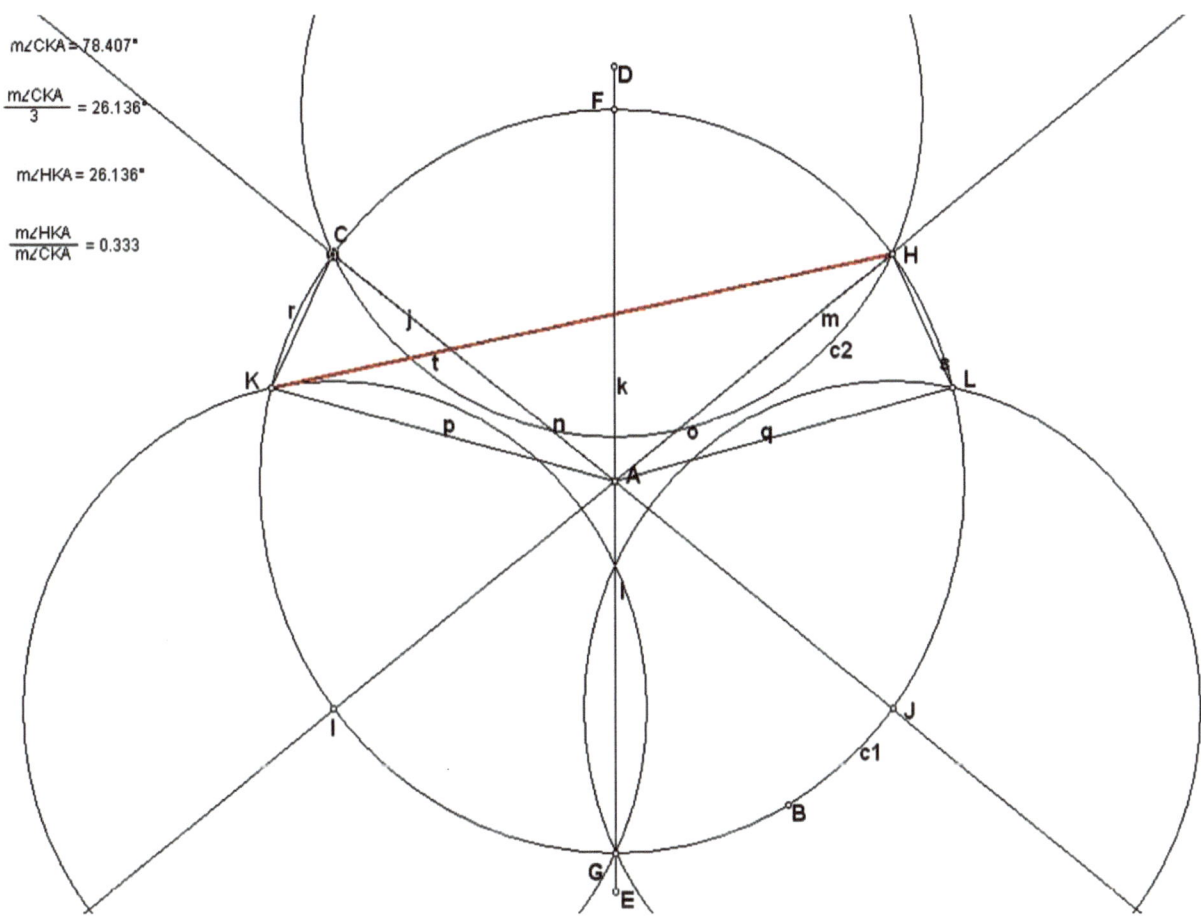

m∠CKA = 78.407°

$\frac{m\angle CKA}{3} = 26.136°$

m∠HKA = 26.136°

$\frac{m\angle HKA}{m\angle CKA} = 0.333$

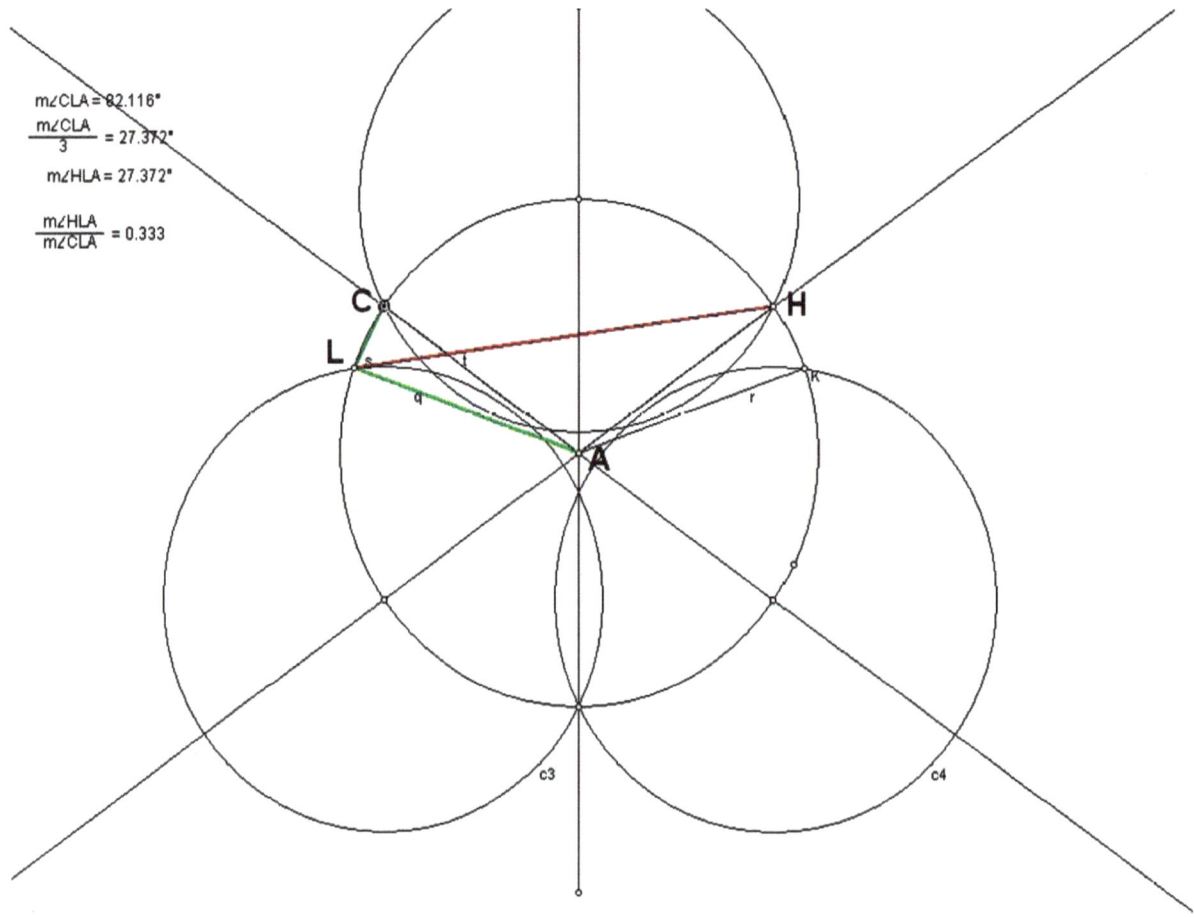

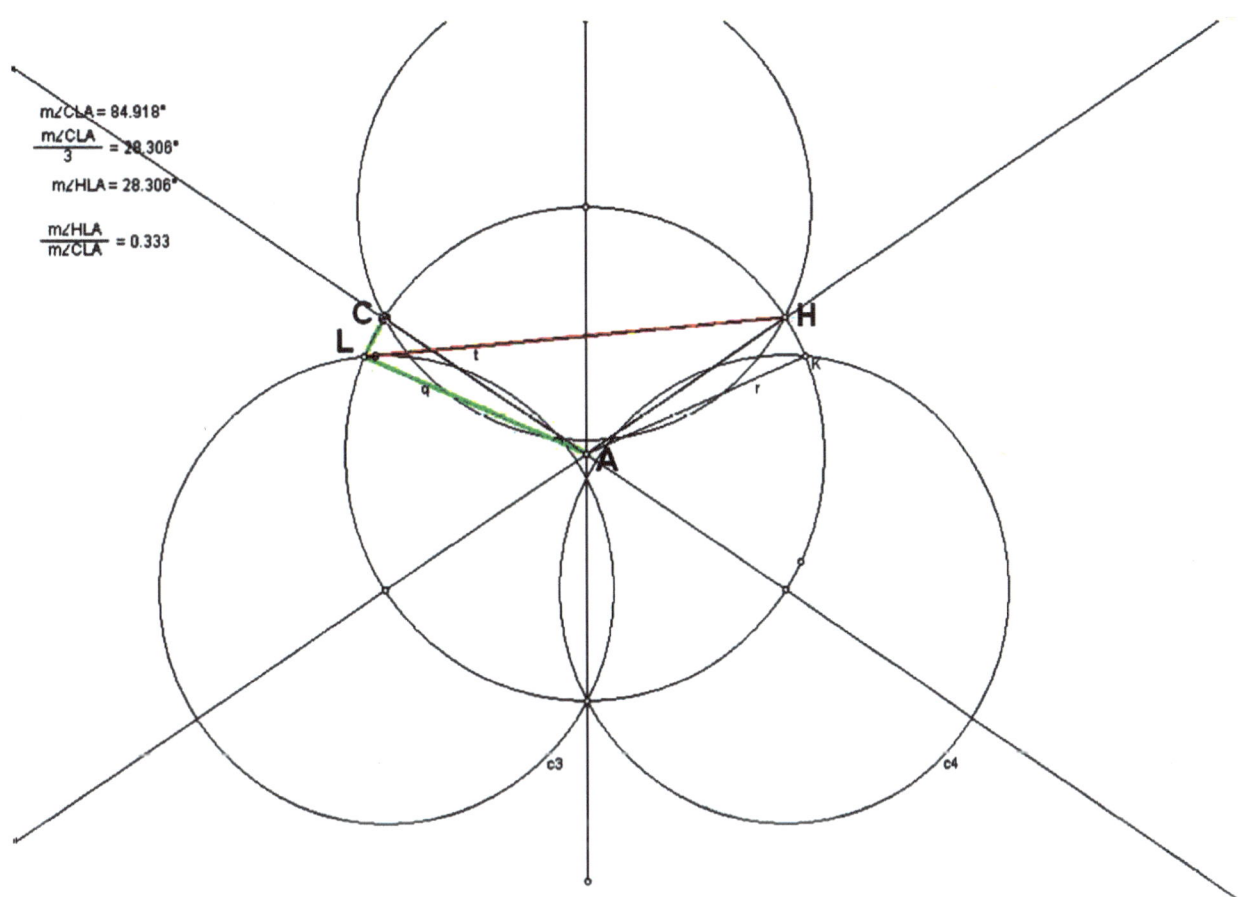

$m\angle CLA = 84.918°$

$\dfrac{m\angle CLA}{3} = 28.306°$

$m\angle HLA = 28.306°$

$\dfrac{m\angle HLA}{m\angle CLA} = 0.333$

V. Summary

The bases of the ART hypothesis is the spatial relationship between coexisting angles. Those relationships can provide solutions to certain Euclidian geometry problems that, up to now, have eluded the community of mathematicians and geometers for over 25 centuries. Specifically, the approach in this paper is to construct a circular figure with inscribed angles of such a geometrical relationship that the bisector of one of the inscribed angles can be used or transferred and transformed to trisect an associated angle. This indirect approach not only addresses an obscure problem in plane geometry, but it also raises a more fundamental question of how we have approached certain 'unsolvable' problems in the past—both from a mathematical and a geometrical point of view. If the ART hypothesis proves correct then perhaps we may need to revisit other impossible problems not only from outside the box, but perhaps, from within the box as well.

VI. References

Bogomolny, A. "Angle Trisection." http://www.cut-the-knot.org/pythagoras/archi.shtml.

Bogomolny, A. "Angle Trisection by Hippocrates." http://www.cut-the-knot.org/Curriculum/Geometry/Hippocrates.html.

Bold, B. "The Problem of Trisecting an Angle." Ch. 5 in *Famous Problems of Geometry and How to Solve Them.* New York: Dover, pp. 33-37, 1982.

Chen, Tzer-lin Proof of the impossibility of trisecting an angle with Euclidean tools, *Math. Mag.* **39** (1966), 239-241.

Conway, J. H. and Guy, R. K. *The Book of Numbers.* New York: Springer-Verlag, pp. 190-191, 1996.

Courant, R. , and Robbins, H., *What is Mathematics?*, Oxford University Press, 1996.

Courant, R. and Robbins, H. "Trisecting the Angle." §3.3.3 in *What Is Mathematics?: An Elementary Approach to Ideas and Methods, 2nd ed.* Oxford, England: Oxford University Press, pp. 137-138, 1996.

Coxeter, H. S.M. "Angle Trisection." §2.2 in *Introduction to Geometry, 2nd ed.* New York: Wiley, p. 28, 1969.

Delattre, J and Bkouche, R, Why ruler and compass?, in *History of Mathematics : History of Problems* (Paris, 1997), 89-113.

Dixon, R. *Mathographics.* New York: Dover, pp. 50-51, 1991.

Dorrie, H., *100 Great Problems Of Elementary Mathematics*, Dover Publications, NY, 1965.

Dörrie, H. "Trisection of an Angle." §36 in *100 Great Problems of Elementary Mathematics: Their History and Solutions.* New York: Dover, pp. 172-177, 1965.

Dudley, U. *The Trisectors.* Washington, DC: Math. Assoc. Amer., 1994.

Dunham , W., *The Mathematical Universe*, John Wiley & Sons, NY, 1994.

Geometry Center. "Angle Trisection." http://www.geom.umn.edu:80/docs/forum/angtri/.

Heath, T L, *A history of Greek mathematics* **I** (Oxford, 1931).

Honsberger, R. *More Mathematical Morsels.* Washington, DC: Math. Assoc. Amer., pp. 25-26, 1991.

Klein, F. "The Delian Problem and the Trisection of the Angle." Ch. 2 in "Famous Problems of Elementary Geometry: The Duplication of the Cube, the Trisection of the Angle, and the Quadrature of the Circle." In *Famous Problems and Other Monographs.* New York: Chelsea, pp. 13-15, 1980.

Loy, J. "Trisection of an Angle." http://www.jimloy.com/geometry/trisect.htm.

Ogilvy, C. S. "Solution to Problem E 1153." *Amer. Math. Monthly* **62**, 584, 1955. Ogilvy, C. S. "Angle Trisection." *Excursions in Geometry.* New York: Dover, pp. 135-141, 1990.

Scudder, H. T. "How to Trisect and Angle with a Carpenter's Square." *Amer. Math. Monthly* **35**, 250-251, 1928.

Steinhaus, H. *Mathematical Snapshots, 3rd ed.* New York: Dover, 1999.

Tabak, J. *The History of Mathematics–Geometry, The Language of Space and Form.* New York: Facts on File (InfoBase publishing), 2004.

Thomas, I, *Selections illustrating the history of Greek mathematics : Vol. 1 (From Thales to Euclid)* (London, 1967).

Thomas, I, *Greek mathematical works* (London, 1939).

Wantzel, M. L. "Recherches sur les moyens de reconnaître si un Problème de Géométrie peut se résoudre avec la règle et le compas." *J. Math. pures appliq.* **1**, 366-372, 1836.

Wazewski, T. *Ann. Soc. Polonaise Math.* **18**, 164, 1945.

Wells, D. *The Penguin Dictionary of Curious and Interesting Geometry.* London: Penguin, p. 25, 1991.

Yates, R. C. *The Trisection Problem.* Reston, VA: National Council of Teachers of Mathematics, 1971.

www.ingramcontent.com/pod-product-compliance
Lightning Source LLC
Chambersburg PA
CBHW041301180526
45172CB00003B/918